I0070739

Successful Contractor Management

In Manufacturing Industries

Successful Contractor

Management

In Manufacturing Industries

Bernard Watson

Successful Contractor Management in Manufacturing Industries
Copyright © 2018 by Bernard Watson. All Rights Reserved.

No part of this publication may be reproduced, stored in a retrieval system or transmitted in any way by any means, electronic, mechanical, photocopy, recording, or otherwise without the prior permission of the author except as provided by USA copyright law.

The opinions expressed by the author are not necessarily those of SCM LLC.

This book is designed to provide accurate and authoritative information with regard to the subject matter covered. This information is given with the understanding that neither the author nor SCM LLC is engaged in rendering legal, professional advice.

© 2018 SCM LLC
Cover design by Panagiotis Lampridis
Interior design by Ayesghee Publishing

Published in the United States of America
ISBN 13: 978-0692082850
ISBN 10: 0692082859

DEDICATION

This book is dedicated to every contractor fatality across the world.

ACKNOWLEDGEMENTS

To my wife Nadine our and children – Tanerria, Xavier and Jordan – whose support and encouragement enabled me to produce this book;

To my grandfather, Cyrus B. Lockwood, and to my mother, Anna M. Watson, who taught me to work hard, be honest, be loyal, have compassion, and most of all, trust God;

Most importantly, I thank God for the opportunities, knowledge, wisdom and courage to challenge the world and to demand change through this book – Successful Contractor Management in Manufacturing Industries.

CONTENTS

INTRODUCTION

This book is intended to instruct and guide contractors, manufacturing organizations, and anyone who utilize contractors in their prospective businesses. It is designed to help you understand what is expected when contractors are working at your facility. Contractor management training is provided to improve your knowledge and understanding before, during and after contracted work is performed.

In the manufacturing industry, contractors play a vital role. Many companies would not be able to operate without the work of contractors. They work alongside manufactory employees during normal operation, maintenance down days, scheduled outages, emergency breakdowns, nights, weekends, and even holidays usually with one goal in mind – to get the job done. Their success is the success of the organization in which they serve and, ultimately, is one of the main reasons why most manufacturing companies are thriving.

Many manufacturing companies view contractors differently. Some even think contractors are not as important as company employees. Contracted employees sometimes are considered not part of the team. Many industrial employees may not care for contractors like they would a company employee. Because of this, when contractors are involved in incidents, industrial employees may not be as concerned and communicate contractor incidents at many facilities simply as, "Oh, it was just a contractor that got hurt. No big deal." There are still many companies that do not even track or trend contractor incidents or provide any type of resource or feedback to help. This mentality has shifted our contractor work force in the wrong direction. Contractor safety, reliability, effectiveness, efficiency, quality, and cost are all depending on the relationship between the contracting company and every team member at the manufacturing facility.

This book is not just for managers and supervisors who work with contractors. Every team member at your facility needs to know, how to carry out their part and what's expected of them when contractors are working in their area. Understanding the importance of supporting contractors on all levels is the key to completing every job in a safe, efficient, and time sensitive method.

While manufacturing companies are focusing on employees' safety, high product quality, speed to the market, and superior customer satisfaction, the importance of contractor interaction has decreased. In all actuality, successful contractor management is a key component to complete all goals listed above.

This book consists of proven contractor processes, procedures, forms, permits, activities, concepts, standards, regulations and requirements needed to manage your contractors successfully. For more in depth training and seminars, please contact: SCM, LLC

FROM THE AUTHOR

…Planting the Seeds of Commitment

My life started off with the fundamentals of hard work at a very young age. My mother and father divorced when I was 2 years old. My mother moved from downtown Charleston, South Carolina back to her hometown of Huger, South Carolina – a small rural town about twenty miles outside of Charleston – to live with her parents back in 1975.

My grandparents, Reverend Cyrus B. Lockwood and Alma Lockwood, lived in a five bedroom home that sat on ten acres of land. That is where I grew up. While Rev. Cyrus Lockwood was his official name, he was commonly known as "Papa" at home. As I think back on growing up in Huger with Papa, the one thing that remains consistent is HARD WORK. He raised many of his grandchildren with the same work ethic. Many of my cousins – Tyrone, Terry, Stevie, Donald, Terrence and Cal – just to name a few – often worked from sunup until sundown. Every season brought on a different type of work – from cutting firewood in the winter, planting crops in the spring and processing those crops in the summer. We sold what we grew and, by Fall each year, we were stocking up firewood, animal food, and food for the family. During my childhood in the late 1970s and early 1980s, Papa taught me many things before he passed away in 1985. I would always remember him saying "always work, work hard, work smart, and plant seeds in life".

I idolized my grandfather, I followed him everywhere. He was well-known throughout the community and respected as a knowledgeable and humble man. Most of all, I remember him for being very loving and one who truly loved God. Being in his presence and listening to his conversation with everyone who crossed his path empowered me to realize that everyone matters, no matter who they were. He would say, "Your most important life lesson will come for someone you did not expect, so give everyone your time". Because of him, I am the man I am today. He planted a seed many years ago and today, through me, it blooms.

FROM THE AUTHOR

…*Working as a Contractor*

I began my career in the manufacturing industry as a contractor at a chemical plant in my hometown of Huger, South Carolina. I worked for a large construction company as electrical helper. I worked with a Journeymen electrician for fifteen months. As an electrical contractor, safety was paramount. Working with electricity was a high risk job in itself. When you factor in the environment (a chemical plant) and its potential hazards, the chances of being hurt, injured or worse doubles.

During this period in my life, I knew I just could not rely on hard work alone. I had to work smart, effectively, and most of all, SAFELY. As contractor, you rarely look at the big picture or the overall project or process. Normally, it is the day-to-day activities or the job you were assigned to complete that concerns you most. Contractors work hundreds of individual projects, sometimes not knowing if it would be their last project for the week or for the entire job.

When I worked as a contractor, it was customary to meet each morning for what we called a "tailgate meeting". During this meeting, our foreman and the project manager would be in attendance, discussing the work that needed to be completed that day. I learned quickly to ask questions and to inquire about schedules. I challenged myself to ask a question every morning at these meetings and I did. At first, my questions were mostly about safety. As I became more accustomed to the process, I would inquire about schedules, then timelines and cost. Asking questions worked to my advantage. It showed that I cared about my safety and the safety of others, that I had educated myself about the project, and it caused my foreman and the project manager to take notice of my concern.

I always believed if you were invited to a meeting, you should weigh in and give your opinion on the matter. No one should just be a spectator at these types of meetings, saying nothing. We all had the opportunity to speak and an opportunity to add value. Many of my co-workers began to ask me questions that they wanted to know the answer to and I would bring it up in the meeting for them. I always believe that everyone's time is valuable. It is the one thing you cannot get back so make it count. Before leaving that particular assignment, I was leading these meetings for twenty-five contractors. After working safely for a year and a half, seeing it to completion, I was laid off with a better understanding of a contractor's role in construction.

My next contractor job was with a door installation company. I worked in several capacities – as a welder, an electrician, and a maintenance tech – on various job sites but, primarily at the local Air Force base. We serviced and installed hanger bay doors and loading docks. On this job, I would work until the job was done, twelve-hour days, nights and weekends. I did it all. I was dedicated and thought my hard work would shine through. Unfortunately, this time, my hard work backfired on me. Simply because I worked harder, I was given more while everyone else was still doing the same amount of work. This went on for about six months until I became burnt out, causing me to resign. Before I did, however, I realized one of my greatest life lessons – to put yourself first. Care about your health, your body, and your life above the job, the client, or the company you are working for because once it's gone, you can never get it back.

When you do things, do not let selfishness or pride be your guide. Instead, be humble and give more honor to others than yourselves.

Philippians 2:3 (NCV)

First Things First...

"As employers seek new ways to make the employment relationship more flexible, they have increasingly relied on a variety of arrangements popularly known as 'contingent work.' The use of independent contractors and part-time, temporary, seasonal, and leased workers has expanded tremendously in recent years."

– Department of Labor

What is a Contractor?

A contractor is an individual or organization performing work for a company but not actually employed by the company. There are several other terms used to describe a contractor – "independent contractor", "subcontractor", "contingent worker", and "hired labor" just to name a few. A contractor can perform a variety of services in a manufacturing environment – ranging from construction to ground maintenance to food service. Contractors may come on site to clean windows, test equipment, service equipment, repair vehicles, or work on new construction projects.

A professional contractor should also have an understanding of his or her limitations. A good contractor understands that the success of the project depends on his or her ability to hire the right people and follow the wishes of the client.

In a different sense, a contractor could also be anyone who agrees to perform work for a fee. This occurs frequently in businesses that cannot afford to assign or hire a new employee to perform a specific job. The job itself may not be long-term enough to justify the expenses of a new hire, or the wages may not be sufficient for established employees.

All contracting companies must be licensed by an examining board or your state before they can work on your site.

The Law

Each state has its own set of mandates regarding contracted labor. On a federal level, Congress has enacted numerous laws geared toward protecting independent contractors. Below are a few of these federally enacted regulations:

- **The Health and Safety at Work Act 1974 (HASAWA)** was put in place to encourage employers to adopt a high standard of the protecting the "health, safety, and welfare" of all employees, including contracted workers. In addition, the Act addresses the responsibilities related to employees and how they should conduct business in a way that does not expose non-employees (which includes contractors) to unnecessary health and safety risks.

- **The Management of Health and Safety Work Regulations (1999)** reinforces HASAWA, requiring that, when contractors are working on an employer's premises, they be provided with information and instructions on the relevant risks to their health and safety which are specific to the premises where the contractors will be working or to the specific activities to be carried out.

- **The National Labor Relations Act** is responsible for creating a standard to avoid the misclassification of independent contractors. This important labor and employment statute helps to determine, along with the Taft-Hartley Act of 1947 (also known as The Labor Management Relations Act), whether an actual independent contractor relationship exists.

A Personal Experience in…
Contractor Coordinating

Contractor coordination consists of several critical requirements that could spans across different departments within a facility. For example, contractors' insurance requirements and legal contracts are normally managed by the procurement department. Safety requirements are handled by your safety department while day-to-day contractor expectations are handled by user departments. The importance of a key person – a Contractor Coordinator – to manage all of these requirements is crucial for your contractors and proven to be a best practice in contractor management. I have found that there are still many manufacturing facilities that do not have a contractor coordinator in place. My research has also shown me that facilities without a contractor coordinator are being plagued with two main issues – (1) important information not shared (effectively or efficiently) or not being made crystal clear with contractors and (2) efforts being duplicated. These deficiencies can lead to a compromise in safety, wasted man hours, and lost revenue. One person primarily focused on contractor coordination will streamline your contractor process by steering your current contractors in the right direction and bringing new contractors on site with the right information in the lease amount of time.

Within my facility, our contractors are family and out concern for their well-being is second nature. In fact, my own mother and many of my biological family members have been or are currently contractors at the facility where I work. Their safety on and off the job is very important to me. As contractor coordinator, it is my job to help them succeed on every task they perform. I wish to accomplish this by proactive planning, safe thinking and over communicating.

Contractor Selection and Approval Process

The contractor selection is one of the most important decisions for an employer. Depending on the size of your project and time frame in which it needs to be completed, the relationship with your contractors can develop into a long lasting one. A thorough selection process ensures the best and most qualified candidates are chosen. It could mean the difference between a successful project and a very stressful failed attempt at one. It is imperative to hire a contractor that has already completed similar projects to those in the manufacturing industry. Check qualifications, references, and perform background checks. Consider the ease of communication with the contractor and the methods used. Check references. Perform backgrounds checks. Creating the right relationship is an investment in your organization. Therefore, it is in your company's best interest to be thorough in selecting the right contractor. Your due diligence will contribute to your satisfaction with your final contractor selection.

Once you are ready to make your choice, be sure to obtain copies of all necessary documents needed to form a contractor relationship. These documents may vary, based on the requirements of your state; however, the following are the basic requirements for the manufacturing industry:

- **Three Years of Occupational Health and Safety Administration (OHSA) Logs** – The contractor should produce logs that record any serious work-related injuries and illnesses for the past three years. These logs help the organization evaluate the contractor's commitment to the safety of a workplace and their understanding of industry hazards. They are also used to prevent future workplace injuries and illnesses by putting the proper worker protections in place.

- **Contractor's License** – This license is issued by the contractor's state licensing board, based on the contractor's trade. Be sure that the license is valid.

- **Insurance Coverage** – At minimum, the contractor should possess the following coverage at the limits required by their state licensing board: worker's compensation, general liability, and automobile liability. A certificate of insurance should be furnished to the company prior to the contractor commencing any work

- **IRS Form W-9** – This form, according to the IRS, is used to provide a contractor's correct tax identification number (TIN) to your organization for the purpose of filing an information return to report income paid to the contractor.

- **Independent Contractors Agreement** – This important document outlines the services to be performed by the contractor. It also includes information regarding the independent contractor relationship and responsibilities, how the contractor will be paid, and other pertinent information. This agreement should be signed by an official representative for both the company and the contractor.

- **Contractor Profile Sheet** – This collection of information regarding the contractor is mostly for recordkeeping purposes. It will include demographic information (address, phone number, email address, website, and tax identification number) as well as other information regarding the contractor's licenses, certifications, training, etc.

All of this necessary information and accompanied documents should be collected and maintained for all contractors. Companies collect and store this information using a method most advantageous to the organization. The data regarding contracted help (both new and established) should also be stored in a way that is easily manageable. Some companies prefer using paper files while others opt for using a web-based data management system. A Flow chart with your company's approval process for new contractors should be trained on by all company team members and posted throughout your facility.

Once your contractor is approved to perform work at your facility, use this opportunity to build a long-term relationship with the contracting company as well as any of the subcontractors. Such relationship will be beneficial to all parties involved. Your company will have an established group of trusted contractors to call on as the need arises and the contractor will gain a customer for life who will also refer them to others.

Careful planning puts you ahead in the long run;
hurry and scurry puts you further behind.
PROVERBS 21:5 (MSG)

Web-Based Contractor Data Management System

Maintaining pertinent contractor information is essential. Every company should have a system, or database, in place for easy storage and access of such information. The type of system used is based on the preference of the company. All databases got their start as paper-based – documents arranged and filed in file cabinets. Wet signatures were required and paperwork was exchanged by manually – by hand, through mail, and/or via fax. Maintaining the safety, insurance, quality and regulatory information for every contractor can eventually impose a strain on internal resources. Fortunately, improvements in technology have made this process a bit more streamlined. A web-based data management system streamlines the process, saving time and money, and improving safety standards. Documents can be signed and forwarded via electronic means. Storage shifts from file folders and cabinets to file space on a computer that can be backed up, restored, and easily retrieved. A web-based data management system also gives manufacturing companies the ability to measure key performance metrics while promoting transparency, clear communication and sustainable operations. Such systems are equipped to identify contractors who meet both manufacturer and regulatory standards. Companies that employ a web-based data management system are able to drive sustainable performance improvements while lowering costs and strengthening relationships with their contractors, as well as track your company's information and receive notifications of upcoming expiration or changes in requirements.

It would be ideal for all companies that work with contractors to utilize a web-based data management system. Doing so will help to standardize contractor management across multiple sites and geographic regions. These systems are designed to ease demands on time and personnel by eliminating duplicated processes, facilitating communication, and promoting transparency. Whether you are a manufacturing company or a contractor, time is money. Web-based data management system help streamline your company's data management system, resulting in lower incident rates and higher compliance numbers. A web-based data management system provides tools, like training and safety programs, that allow your company to be proactive, focusing on continuous improvement, so you can avoid spending time playing catch-up.

A Personal Experience in…
Contractor Networking

In November 2011, I attended the 2011 ISN Annual Users Conference in Dallas, Texas and had the opportunity to share my experience in contract management. There were about 3000 attendees from all over the country – all there to learn, share best practices, and network in the name of contractor management. During the conference, I had the privilege of presenting to over 300 attendees. At this point in my career, I had been presenting to approximately 100 supervisors and managers on a monthly basis for about five years. This group was my largest by far.

My presentation was entitled "Managing Day to Day Responsibilities with Contractors" and it was scheduled for 2:50pm on the first day of the two-day conference. I was well-prepared, having created my presentation slides months ago. I knew the presentation material really well primarily because I created most of it and had presented it in the past. At 2:30pm, I started to get excited and before you knew it, it was my time. I got on stage, started to speak, and, just like that, it was over. I nailed it.

After my presentation, I had nearly 30 people waiting to talk to me about my presentation. They were all extremely impressed with how we handle certain things at our facility. One gentleman approached me and said something that changed my view of what I did and how well put together it was. He said, "Your knowledge, compassion, and love for what you're doing shows through your presentations", I said, thank you, shook his hand and walked away. At the end of the first day, all of the presenters were allowed to meet and converse with the conference's keynote speaker, the 22nd Secretary Of Defense, Dr. Robert Gates, who, in my opinion, is one of the greatest to serve in that capacity. This event sealed the deal for me; I had to put what I knew on paper and share it with the world.

Secretary Of Defense Dr. Robert Gates and me

November 3, 2011, ISN Annual Users Conference, Dallas TX

Contract Workers' Training

In the manufacturing industry, all contract workers should possess a competent level of skill and training in their respective trades. Every trade requires some sort of continuing education or annual training requirements as a condition for proper licensing and certification. In addition, contract workers should complete annual basic training in the following areas prior to working at your facility:

- Emergency & Fire Prevention Plans

- Noise Exposure

- Personal Protective Equipment (PPE)

- Electrical Safety

- Respiratory Protection

- Fall Protection

- Hazard Communication

Another important area in which you want to ensure contract workers have proper training surrounds operating mobile equipment. Be certain that anyone operating any type of equipment at your facility is properly certified to do so. Your company should require that all training be documented and verified before any work can take place. This is for the protection of your organization as well as that of the contractor.

Contractor Communication

You have heard it before – communication is key. Effective communication, or lack of, can make or break the relationship between your company and your contractor. Establishing a communication plan with your contracted workers should be one of the tasks Knowing that communications will be flowing freely throughout the project tends to lessen project coordinator anxiety and makes getting through the project process a little easier.

Save yourself communication headaches by interviewing your contractor and accessing the communication style used before committing to that particular contractor. A company's communication style often is revealed during your first contact. Red flags during the initial interview could be an indication of future issues or concerns. Here are some important points to consider when communicating with a potential contractor –

Did the company respond to your initial call in a timely manner?

Do you reach a person or a voicemail?

Did the contractor listen to me? Do you feel heard?

Is the contractor experienced in various forms of communication?

Ask similar questions when checking a contractor's references. Find what experience references have had regarding communication with the contractor. In addition, while evaluating the contractor's communication style, discuss your organization's communication styles as well. This helps to ensure compatibility. If you prefer phone calls over emails, discuss that. If you hardly answer the phone but respond quickly to text messages, let that be known. Keep in mind that communication is directly related to establishing and maintaining a successful working relationship.

Developing successful communication is key when working with contract workers. You should make contact with them every day. There are many ways this can be established to ensure responsive results. The following methods of communication are typical within the manufacturing industry:

By Phone:

Establish and maintain an up-to-date phone list of all contract workers as well as direct supervisors and/or members of management from the contracted company. This list will make for easy contact with important personnel at any time during the day. Keep in mind that, for the safety of everyone involved, cell phones should never be used while operating equipment or walking around the facility. Only use cell phones in a safe, designated area.

By Email:

E-mail is an effective method of communication because of its convenience. It can be easily access from our computers, our phones, and our tablets. We can send and receive from anywhere in the world. In the same regard, email can also be a. Because of its convenience, it becomes very easy for us to use email when it may not be the best option for some conversations. When using email to communicate with contract workers, be sure to:

1. **Be clear. Be brief** – Write short, concise email. Long, confusing paragraphs can lead to misinterpretation.

2. **Proofread** – Proofread your message multiple times. If the email deals with touchy subject matter, have a third-party read it over as well. This can give you added perspective on how well it will be received by the intended recipient.

3. **Wait** – Write the email, proofread it, and then sit on it for a little while. Knock off a couple other tasks before you have second look at it. If your emotions were flaring when you wrote the email, a little time can allow these emotions to settle, allowing you to evaluate the message in a different light.

4. **Pick up the phone** – While email can be a convenient communication channel, certain discussions need to be handled over the phone, or if possible, in person. If you notice that the situation is starting to deteriorate, don't send another email, it's time to pick up the phone or arrange a face to face meeting

In-Person Meetings:

Each week at a set day and time, plan a meeting to discuss the progress of the project. This is your opportunity to express any concerns you might have or changes you wish to make. This meeting should be held with the contractor manager.

It is critical that you establish a clear chain of communication and command for the input and distribution of information. All requests for information, change order requests, and directives to and from the Client should be introduced in writing and addressed through proper channels to ensure issues are responded to by the right party without delaying progress. Captured, documented and file all correspondence for every project when complete.

When you talk, you speak of something you already know but, if you listen, you might learn something new.

DALAI LAMA

Contractor Safety Orientation

Contractor Safety Orientation is intended to inform and educate contractors of a facility's expectation regarding health, security and environmental concerns. Every contractor, subcontractor, vendor, supplier and contingent employee should receive a formal orientation, as a condition of work authorization, before commencing work at a manufacturing facility.

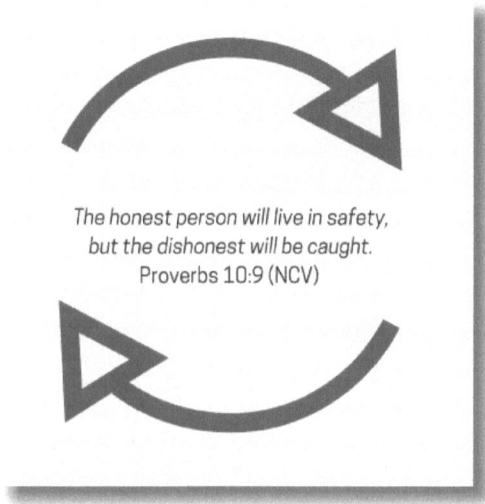

> The honest person will live in safety,
> but the dishonest will be caught.
> Proverbs 10:9 (NCV)

All contractors will be responsible for following these policies and procedures clearly. All contractors are supposed to have the proper orientation in both safety and environmental matters by their own employer upon hire. In order to verify that the contractor has, in fact, had orientation in the required safety and environmental topics, and to assess their level of understanding and retention of these topics, contractors should also be required to attend safety and environmental training specific to your facility followed by a post-evaluation assessment. Only contractors who receive a perfect score on this assessment should be allowed inside.

Your facility orientation should cover your its policies, programs, procedures, and rules. Any manufacturing facility can employ a number of rules that cover every aspect of operations. The following list includes some common expectations and requirements for contractors within a manufacturing facility.

This list is not all inclusive, by any means, but a helpful tool for evaluating what is important in protecting your facility, your employees, and those you contract to work within your organization.

Common Expectations for Contractors within the Manufacturing Industry:

- ***Movement on Site*** – Vehicle and pedestrian movements on site must be in accordance with local traffic and restricted to those areas agreed with the facility;

- ***Competent Persons*** – Contracted workers must be competent to undertake the work specified and be able to operate any necessary equipment safely and without risk;

- ***Risk Assessments*** – Any activities that may pose a significant risk to the health and safety of people, within the facility, or to your product must be formally assessed. Those most likely to be affected by these potential risks should be informed and safety measures should be taken;

- ***Tools and Equipment*** – All tools and equipment used to perform daily tasks should be in good working condition and fit for its intended purpose;

- ***Portable Electric Equipment*** – Said equipment must possess a current and up-to-date test certificate. Voltage requirements should not exceed 100 volts unless protected by a suitable residual current circuit breaker;

- ***Personal Protective Equipment (PPE)*** – For certain tasks or in specified areas the wearing of suitable PPE shall be strictly enforced;

- ***Hygiene*** – Any applicable hygiene rules will be strictly enforced;

- ***Fire*** – Notices are displayed throughout the site; you must be aware of site procedures what action to take if the fire alarm is sounded or if you discover a fire;

- ***Accidents*** – All accidents must be reported immediately. A site contact and the contact information should be provided during orientation along with proper procedures for dealing with various accidental situations;

- *Smoking* – If smoking is permitted within the facility, a smoking policy will be in place and this policy must be strictly observed.

- *Alcohol and Controlled Substances* – Due to the nature of the industry, the consumption of alcohol on site is often strictly prohibited. Moreover, anyone found to be under the influence of alcohol or other controlled substance should be removed from site and not permitted to return.

More specifically, items listed below may be included also:

- Incident investigation
- Accident prevention signs and tags
- Hand safety
- Back injury prevention
- Behavioral safety
- First aid/CPR/AED
- Confined Space
- Lockout/Tag out
- Hazardous communication
- Intervention (training)
- Walking working surfaces
- Job Safety and Environmental Analysis
- Respiratory
- Prevention of workplace violence
- Fall protection: introduction
- Permitting
- Emergency evacuation

- Environmental
- Excavation - trenching and shoring
- Occupational Health

Contractor Safety Orientation should be considered as annual training for all contractors working at your facility. These orientations are verified by individual contractor sign in logs and formal audits.

Note: Safety orientations provide an awareness level understanding of work hazards and may not fulfill training requirements, as prescribed by Federal, State regulations and industry standards.

2

On-Site Procedures

"...to the extent that a subcontractor of any tier agrees to perform any part of the contract, he also assumes responsibility for complying with the standards in this part with respect to that part...With respect to subcontracted work, the prime contractor and any subcontractor or subcontractors shall be deemed to have joint responsibility."

– Occupational Safety and Health Administration (OSHA)

Quoting On-Site Jobs

Quoting and estimating projects is a cost-related process involving the contractor and the owner/client. Requests for quotations (RFQ) are most commonly used in the business environment. Information like safety expectations, rental equipment, payment terms, quality level, project drawings and project timeline are, more than likely, requested during this process. Project quotes should always be done in a written format or electronically, never verbal. Always have at least 3 different companies to submit a quote on your project. Most projects are typically classified in two ways: In-House and Permitted.

All projects that are located inside an existing building are considered an **In-House Project**. In-house projects normally will not require permits or engineered-stamped drawings prior to starting. All in-house projects should meet the following criteria (at minimum) before work can begin:

1. In-house projects must be approved by your company-employed engineer or area maintenance supervisor;

2. In-house projects must be built to OSHA, IBC, and your company standards;

3. Any contractors used for in-house projects must view existing drawings prior to starting the project;

4. Contractors used for in-house projects must supply a drawing or sketch of all modifications to your company-employed engineer or area supervisor prior to starting the project;

5. Once the work is completed, the project contractor must follow your company procedure for project close out.

All projects that require a foundational slab outside of an existing building structure is classified as a **Permitted Project**.

1. All permitted projects must be professionally engineered.

2. Most local counties require stamped drawings done by a certified Professional Engineer (PE) when applying for permits.

3. Your engineering drawing package should include foundational, structural, and electrical drawings.

4. Your engineering drawing package should subsequently be submitted to your company's environmental department for storm drain evaluation.

5. The Company's general contractor is responsible for applying for all permits. The general contractor is also responsible for picking up the permits after approval.

6. Upon environmental approval, your drawings should be submitted to

your local county office for approval.

7. The local county office must issue a permit before any work can be performed.

8. During project construction the general contractor will conduct pre-inspections and schedule phase inspections.

9. All inspections must be completed by an authorized inspector.

10. Once the final inspections are done and the job is 100% complete, contractor will release all final documents (drawings, permits, schedules, specification sheets, invoices, and etc.) to your company's engineering department for proper filing.

When quoting both types of projects, scope must be crystal clear across the board. Using a project quoting sheet will insure this is achieved. (See the attached sample).

SCM, LLC

New Project Quote Sample

Start Date _____ Project Name _____

Project Description _____
_____

Project safety factors _____
_____

Circle One: Repair   Upgrade   New Installation   Safety

Area _____

Site contact _____

W/O or CEA# _____

Contractor #1 _____ Quote _____

Contractor #2 _____ Quote _____

Contractor #3 _____ Quote _____

Equipment supplier _____ Delivery date _____

Estimated Material Cost _____

Equipment Rental _____

Associated Drawings _____

Sketches Available _____

Stamped Drawings Required _____

County Permits Required _____

Shutdown Installation required _____

Engineering Vendor _____

Estimated Completion Date _____

Work Order Generation

Upon quoting completion and selected contractor is awarded the project, the next step will be entering a work order for your contractor to start. A work order or job order (sometimes refer to as job ticket or work ticket) is an order received by the contractor from a client. A work order may be for products or services. Work orders should always be in place before any contracted work is performed on site. For situations that require immediate attention—including situations that pose a life safety/health hazard or cause major property damage, such as fires, floods, and emergency production equipment breakdowns – a manager's verbal work request will suffice.

Most companies utilize electronic maintenance software to manage and track contractor work on site. Electronic maintenance software is a truly customizable solution built to not only manage your Contractors and projects, but help you gain real operational advantage through powerful tools. With maintenance software, all of your assets, work orders, purchase orders, schedules, warranties, and inventory are located and readily available in one place. No matter how many locations you have, no matter how many projects—all data is instantly shared and accessible across your entire organization.

When generating a new work order, be sure to have the following contractor information available:

1. Contractor contact name, phone number, and email address
2. Location where services are needed (including building, floor, and room number)
3. Detailed description of the service required
4. Project total cost, including labor, supplies, rental equipment, etc.

Once submitted, each request will be logged into the work order system. Normally, all work order request must be pre-approved by your supervisor and manager. Upon approval, the contractor listed in the request will receive an email confirmation. The work order is now active and the contractor is good to start the project.

When the project is complete, contractors should submit a copy of the work order along with their invoice form that contains the customer information, describes the work performed and lists charges for material / labor. This can be submitted to the customer as an invoice.

By using electronic maintenance software, your company is able to eliminate bottlenecking (stalls in production) and increase communication. Work order systems provide a much quicker process with contracted service, supporting project transparency, traceability and accountability.

Below are other concerns often addressed in most work order systems:

- Recording and tracking projects and work requests;

- Capturing labor, parts, and material costs by customer, location, department, cost-center, and other user-definable criteria;

- Point-and-click planning and scheduling tools;

- User-definable email notification capabilities – automatic or on-demand;

- Assigning the job procedures of multiple contractors and multiple projects to work orders;

- User-definable cause, failure, and remedy codes specific to asset types;

- Tracking job time, response times, machine downtimes, and other important metrics;

- Assigning work orders to projects and viewing project histories;

- Uploading manuals, drawings, or any rich media documents and relating them to work orders;

- Submitting material requests as a standalone request or as an addition to a work order;

- Access to mobile devices;

- Associating project cost and tracking cash flow;

- Invoicing automatically;

- Generating supervisor and manager approval for new work orders

... From My Experience

In my current role in contractor management for a large steel manufacturer, I have had the pleasure of working with contractors on a daily basis. I have learned that most contractors recognize that they play an important role and come prepared with the willingness to perform the tasks required of them. Providing them with trusted processes designed specifically for your facility is essential and necessary in order for contractors to complete their jobs safely, efficiently, and in a timely manner. Ultimately, doing so benefits all parties involved. I have found that once your contractors take ownership in the processes laid out for them, they will utilize them before you ask, even away from your site.

I remember performing a safety audit at manufacturing facility in the Southeast. During the inspection, I recognized a contractor that worked at my current plant location and I stopped him for a quick chat. He was on a 30-day temporary assignment at this facility. During our chat, I asked him if he was enjoying his current assignment and he replied that it was a bit different than what he was used to. "What's the difference?" I asked. With his response, he pulled out a form from his work gang box and handed it to me. As I looked it over, I realized it was a contractor work permit from our facility. He had scratched our facility name off of the permit, which only left *WORK PERMIT* on the top of the form. "They don't have work permits here for contractor use so I'm using the work permit and procedure from your facility," he said. I was so pleased to see that this contractor was using what he considered to be a very important tool for his safety without being asked to do so. He also mentioned that he recognized the value in the work permits and that he will use them on all jobs he performs at any facility he is assigned to.

I was very elated that he trusted in the work permit process and would use it wherever he works. I truly believe doing the right thing when no one is looking is a true testament of personal responsibility at best.

3

Permit Types

To acquire wisdom is to love oneself; people who cherish understanding will prosper.

Proverbs 19:8

Why Work Permits?

In the manufacturing industry, safety is of the utmost importance. There are several ways companies work to ensure the safety of its employees and of its contractors. One process – the contractor work permit process – will be one of the most used processed within your facility when it comes to safety. This process provides a road map to completing every project safely. When used correctly, this process ensures everyone involved in the job is aware and informed that contractors are working in the area. This very important process has been deemed a best practice and ensures that all projects are done according to local, state, and federal government requirements. This chapter will give you some insight on several types of permits and the procedures for obtaining required permits for various types of projects.

Contractor Work Permit & Procedures

Various types of work permits are required while performing any type of tasks within the manufacturing industry. A work permit is required for all contracted work on site, hot work, excavation, confined space work, electrical, alterations to or overhaul of plant or machinery where mechanical, toxic or electrical hazards may arise.

This permit should be completed by the person requesting the work and the contractor prior to work commencing. A copy of the form must be kept at the work location. The objective of the contractor's safe Work Permit system is to identify hazards associated with a project, develop precautions required to control each hazard identified and notify the correct personnel that contractors are working in the area.

Safe Work Permits will help to prevent damage or injury while contractors are working in and around your facility.

Once a safe work permit system is created, company employees will be trained in its purpose and use, and progress in its implementation monitored. It also helps for the employees to be involved as much as possible in writing the program so they feel as though they own the process, instead of it being a corporate mandate via e-mail from a person no one has met. The employees are also experts in detecting unsafe working conditions. Their help will make your facility as safe as possible.

All contractor safe work permits need to be authorized by the job initiator and the area supervisor before work is started. This usually involves the signature of the shift supervisor at a minimum. Before authorizing, the permit should be reviewed to ensure that all steps have been taken to reduce the risk of accident. The contractor performing the work should also sign the permit. To ensure proper coordination between plant personnel and outside contractors, the contractor's project manager should also sign the permit stating that they know and understand the permit provisions.

The following are examples of the various job types that normally require permits:

1. **Hot Work** - the use of open flame, oxyacetylene burning, tar kettles, etc. and/or the use of portable spark or heat producing equipment in flammable storage or handling areas

2. **Confined Space Entry** – limited or restricted means for entry or exit and is not designed for continuous occupancy

3. **Excavations** – the use of heavy equipment to dig and unearth

4. **Blasting** – the use of explosives

5. Whenever required by an existing company's or contractor's procedures or required in the contractor's project safety plan.

6. Use of internal combustion engines, or vehicles with internal combustion engines inside company buildings

7. Use of company owned equipment, such as personnel lifts, fork trucks, vehicles, etc., by employees of a contractor

Safe Work Permit Procedures

Contractors will normally get any required permits from the facility's safety, maintenance or engineering department. In almost every instance, a contractor will have to obtain a work order number before work can begin. In emergency situation, time may not allow for the issuance of a work order number. In this case, a manager's signature could be a substitute for a work order number. Nevertheless, the following procedure is pretty standard for the issuance of a safe work permit:

1. The contractor will fill out all appropriate information and take permit to the company representative involved with the job.

2. The employee will advise the contractor of any job specifics necessary to perform the assigned work. Both will complete the contractor permit to work. Company employee must walk the job with contractor identifying specific safety precautions and job hazards prior to signing permit.The contractor will take the completed permit to the area

operations supervisor to sign and advise him of work to be done in his area. Area supervisor will sign permit and retain the pink copy as a record of the ongoing job.

3. The contractor will keep and use this permit as a reminder of all safety precautions needed throughout the duration of the entire job. If the job takes longer than one shift or day, a new form is not required; however, the permit should be re-signed by the new area operations supervisor every shift. If the job is not continuous, the contractor must pick up the pink copy of the permit at the end of the work day.

4. Once the job has been completed, the contractor must remove all equipment, locks, and clean up any trash, etc. from the job.

5. Once the cleanup has been completed, have the company employee in charge of the job sign off that all work and cleanup has been completed.

6. Capture all changes via sketches and/or auto-cad drawings for proper engineering document modifications.

7. The Contractor will then return to the area operations supervisor to advise him that the work is completed and he is clear from his area. The area operations supervisor will sign and close the work permit.

8. Take the time to **COMPLETELY** fill out permit to the maintenance and engineering departments. A copy will be presented for proper filing and retain a copy for your company records. Now you may proceed with invoicing the work performed.

Whatever you do, work at it with all
your heart, as working for the lord,
not for men..
Colossians 3:23 (NIV)

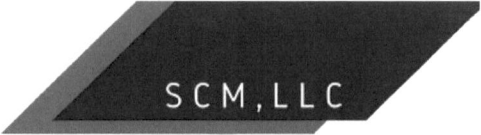

Sample Contractor Safe Work Permit Form

Contractor: _____

Job Order #: _____

Work Description: _____
_____

Date Issued: _____ Time of Issue: _____

Expiration Date: _____

Work Priority (circle one):   Emergency   Urgent   Routine   Shutdown

Special Requirements/Safety Precautions: _____
_____

Type of Permit:
    _____ Hot Work   _____ Electrical
    _____ Confined Space   _____ Crane Inference
    _____ Overhead Cranes   _____ Grating/Floor Plate/Handrail
    _____ Barriers Needed   _____ Hydraulics / Pneumatic
    _____ Hot Molten Metal   _____ Roof Access Permit
    _____ Combustible Gases/Materials   _____ Blue Flag Permit

Area Specific (please Describe): _____
_____

Specific Job Hazards: _____
_____

Lock Out/Tag Out
Equipment to be locked out:_____ Maintenance Contact: _____ Lockbox
Location(s):

Permit Confirmation:
I confirm that the above work is to be undertaken as specified. All site safety requirements
will be met and maintained.

Contractor: _____

I have checked the job with Contractor and identified all area specific hazards. I authorize work to commence.

Employee: _____

Operations Notification: Area Operations Supervisor Notified of work in area.

Permit Closure:
I confirm that the above job is completed. All my employees, lockout devices, barriers, tools, and all trash have been removed from the area. Work under this permit is complete.

Contractor: _____ Date: _____
Employee: _____ Date: _____
Area Supervisor: _____ Date: _____

Hot Work Permit Procedures

Hot work is any type of work that uses an open flame or produces sparks that could start a fire. In order to perform this type of work within an enclosed space, a hot work permit is required. This type of permit provides documented advanced approval for this type of work in a specific area. The following information provides guidance for contractors, subcontractors and other contingent workers who manage, supervise, and perform hot work. These procedures cover provisions to prevent loss of life and property damage from fire and/or explosion as a result of hot work.

Examples of hot work processes include:

1. welding and allied processes – to include, but not limited to, torch cutting, open-flame soldering, and brazing

2. grinding

3. torch-applied roofing

4. any similar applications producing a spark, flame, or heat

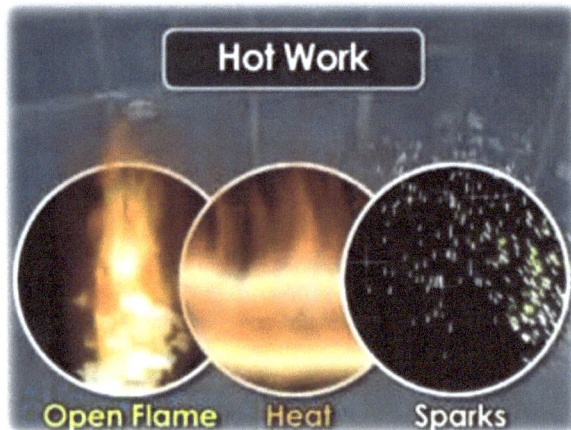

Once it has been determined that a hot work permit is required, a contractor will first establish permissible designated areas for hot work and that all contracted employees involved in the hot work operations understand and

comply with the provisions of this standard. Next, it is necessary to ensure that only approved apparatus, such as torches, manifolds, regulators or pressure reducing valves and acetylene generators, be used by contract employees and advise the facility as to the content of the standard, site-specific flammable materials, hazardous processes or conditions, or potential fire or otherwise hazardous conditions. All of these requirements help to ensure that subcontractors that the responsibilities designated herein for subcontractors and lower tier subcontractors are complied with.

The individual authorizing the permit will be responsible for inspection of areas where hot work is to be performed and for the issuance of permits for such activities.

In addition, this individual will also be responsible for the following:

- determining all site-specific flammable materials, hazardous processes, or other potential fire hazards present or likely to be present in the work location

- determining whether fire protection and extinguishing equipment are properly located at the site

- assigning a fire watch at the site; in the event there are combustible materials that may be affected but cannot be observed by a single fire watch, a second watch is assigned

- ensuring the protection of combustibles from ignition by the following means – (1) moving work to a location free of combustibles, (2) move or safely shield combustibles when work cannot be moved, (3) schedule hot work when operations that could expose combustibles are not being performed, and (4) avoid hot work when the previously mentioned conditions cannot be met.

During hot work, the responsibilities of all partied involved could vary. For example, the individual performing the hot work will be responsible for being properly trained in the safe operation of his or her equipment and for the safe use of the process. Anyone performing hot work should be fully aware of the inherent risks involved with such work and possess an understanding of the emergency procedures in the event of a fire. He or she should be able to handle the equipment safely and use it as described in this procedure so as not to endanger life or property. Before commencing work, this individual performing the hot work should have the approval of the facility's Permit Authorization Individual (PAI) and comply with the requirements of the permit. This individual is also responsible for ceasing all hot work operations if unsafe conditions develop and notifying management, the area supervisor, and/or the PAI for reassessment of the situation.

STEPS FOR PROPER PERMITTING - HOT WORK

- Obtain a hot work permit application

- Complete permit application with all pertinent information

- Gain proper authorization

- Follow all necessary precautions

- Commence and complete required work

- Close hot work permit

The responsibilities of a fire watch during hot work include, but are not limited to:

- being present during hot work operations and remain for a minimum of 30 minutes after completion of hot work in order to detect and extinguish smoldering fires.
- being aware of the inherent hazards of the work site and of the hot work
- ensuring that safe conditions are maintained during hot work operations
- having the authority to stop the hot work operations if unsafe conditions develop
- having fire-extinguishing equipment readily available and shall be trained in its use
- being familiar with the facilities and procedures for sounding an alarm in the event of a fire
- watching for fires in all exposed areas surrounding the hot work operation and try to extinguish them only when the fires are obviously within the capacity of the equipment and fire-fighting skills available
- immediately contacting fire and emergency services when a fire has grown beyond control

Hot Work Areas

There are basically two types of hot work areas: the designated area and the permit-required area. The designated area is a specific area approved for such work, such as a maintenance shop or a detached outside location, and is of noncombustible or fire-resistive construction. It is an area essentially free of combustible and flammable contents and suitably segregated from adjacent areas. These designated areas are generally long-term for facilities in which specific operations are repeatedly performed. A fire watch is not normally required in a designated area. A permit-required area shall be an area that is made fire safe by removing or protecting combustibles from ignition sources. Permit-required areas are generally transient in nature during the performance of varied procedures.

Regardless of the area type, before a hot work permit is issued, it is common practice that the following conditions be verified:

1. Hot work equipment to be used should be in satisfactory operating condition and in good repair

2. Where combustible materials – such as paper clippings, wood shavings, or textile fibers – are on the floor, the floor should be swept clean for a radius of 35 feet minimum. Combustible floors (except wood on concrete) should be kept wet, covered with damp sand, or protected by noncombustible or fire-retardant shields. Where floors have been made wet, personnel operating arc welding or cutting equipment shall be protected from possible shock.

3. All combustibles should be relocated at least 35 feet horizontally away from the work site. If relocation is impractical, combustibles shall be protected with fire-retardant covers or otherwise shielded with metal or fire-retardant guards or curtains. Edges of covers at the floor shall be tight to prevent sparks from going under them, including where several covers overlap when protecting a large pile.

4. Openings or cracks in walls, floors, or ducts within 35 feet of the site shall be tightly covered with fire-retardant or noncombustible material to prevent the passage of sparks to adjacent areas.

5. Conveyor systems that might carry sparks to distant combustibles shall be shielded.

6. If hot work is performed near walls, partitions, ceilings, or roofs of combustible construction, fire-retardant shields or guards shall be provided to prevent ignition.

7. If hot work is performed on a wall, partition, ceiling, or roof, caution should be taken to prevent ignition of combustibles on the other side by relocating combustibles. If it is impractical to relocate combustibles, a fire watch on the opposite side from the work shall be provided.

8. Hot work should not be attempted on a partition, wall, ceiling, or roof that has a combustible covering or insulation, or on walls or partitions of combustible sandwich-type panel construction.

9. Hot work performed on pipes or other metal that is in contact with combustible walls, partitions, ceilings, roofs, or other combustibles should not be undertaken if the work is close enough to cause ignition by conduction.

10. Fully charged and operable fire extinguishers that are appropriate for the type of possible fire shall be available immediately at the work area. If existing hose lines are located within.

11. The hot work area defined by the permit, they shall be connected and ready for service, but shall not be required to be unrolled or charged.

12. If hot work is done in close proximity to a sprinkler head, a wet towel should be laid over the head and then removed at the conclusion of the welding or cutting operation. During hot work, special precautions must be taken to avoid accidental operation of automatic fire detection or suppression systems (for example, special extinguishing systems or sprinklers).

13. Nearby personnel should be suitably protected against heat, sparks, slag, and so on.

14. Based on local conditions, the contractor must determine the length of the period for which the hot work permit is valid; this will also be noted on the hot work permit.

SCM, LLC

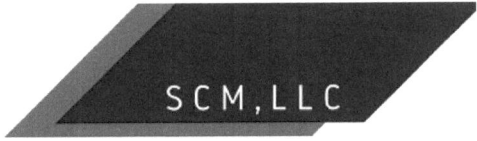

Sample Contractor Hot Work Permit Application

Date: _____   Building: _____

Department: _____   Floor: _____

Work to be Done: _____
_____
_____

Special Precautions: _____
_____
_____

Is a fire watch required: _____

Specific Job Hazards: _____
_____

Permit Confirmation:
The location where this work is to be done has been examined, necessary precautions taken, and permission is granted for this work.

Signed: _____
        Permit Authorizing Individual (PAI)   Date

Final Check:
Work area and all adjacent areas to which sparks and heat might have spread (including floors above and below and on opposite side of walls were inspected 30 minutes after the work was completed and were found fire safe.

Signed: _____
        Permit Authorizing Individual (PAI)   Date

ATTENTION

Before approving any hot work permit, the contractor shall inspect the work area and confirm that precautions have been taken to prevent fire in accordance with NFPA 51 B.

PRECAUTIONS

• Sprinklers in service
• Hot work equipment in good repair

WITHIN 11 M (35 FT) OF WORK
• Floors swept clean of combustibles
• Combustibles floors wet down, covered with damp sa metal, or other shields
• All walls and floor openings covered
• Covers suspended beneath work to collect sparks

WORK ON WALLS OR CEILINGS
• Construction noncombustible and without combustible covering
• Combustibles moved away from opposite side of wall

WORK ON ENCLOSED EQUIPMENT (Tanks, containers, ducts, dust collectors, etc.)
• Equipment cleaned of all combustibles
• Containers purged of flammable vapors

FIRE WATCH
• To be provided during operation and at least 30 minutes after operation
• Supplied with a fully charged and operable fire extinguisher
• Trained in use of equipment and sounding fire alarm

FINAL CHECK
• To be made 30 minutes after completion of any operation unless fire watch is provided

Signed: _____
                    Contractor

Roof Access Permit

When accessing roofs, all contractors must be aware of the safety hazards associated while working on roofs. Any work taking place on or above a roof requires at least the same safety precautions and permits as would be required elsewhere, such as fire watches/extinguishers and hot work permits to perform welding.

Before any contractor is given access to a roof, a roof access permit should be completed and signed by your engineering department.

WARNING

ROOF ACCESS
BY PERMIT ONLY

SCM, LLC

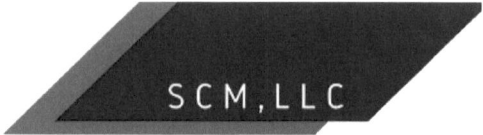

Sample Roof Access Permit Form

1. REQUEST (Roof access Leader):

Name: _____ Job Order Number: _____

Company: _____ Position: _____

Reason for Roof Access: _____

Access Date: _____ Start Time: _____ End Time: _____

2. SITE DETAILS:

Building Name Areas: _____ Area of Roof_____

3. Is additional Fall Protection required? _____ Yes _____ No (Describe below)

_____

4. Is work to be performed that requires a Contractor Work Permit? ____Yes ____ No

b. Additional Hazard Identification: List additional requirements below; _____

_____

5. Additional Requirements: _____

6. APPROVALS:

Signature: _____ Date: _____ Employee No. _____

Engineering

Signature: _____ Date: _____ Employee No. _____

Area Supervisor

7. PERSONNEL IN ROOF ACCESS TEAM:

Name(s): _____ Company(s):_____

Signature(s):_____

Name(s): _____ Company(s):_____

Signature(s):_____

8. WORK COMPLETE AND ROOF CLEAR

Signature: _____ Date: _____ Employee No. _____

Engineering

Signature: _____ Date: _____ Employee No. _____

Area Supervisor

Confined Spaces Permit

An area that has limited access and a restricted means of entry or exit is considered a confined space. While a confined space is large enough for a person to enter, it is not intended for continuous human occupancy. A permit-required confined space meets the above criteria and present hazards or materials that can entrap, engulf, or asphyxiate. Many workplaces contain spaces that are considered confined because the configurations of these spaces hinder the activities of contractors who must enter, work in, and exit them. Confined spaces include, but are not limited to, underground vaults, tanks, storage bins, manholes, pits, silos, process vessels, and pipelines.

Coordination of entry activities shall be sufficient to ensure that the hazards present in the space and the hazards that could be generated by the work activities in the space are clearly understood by all parties involved. The contractor is responsible for contacting your site's safety representative to obtain a copy of the Confined Space Hazard Assessment.

All contractors who perform confined space entry operations are required to be trained in all applicable OSHA Confined Space Entry Standards. For example, entry supervisor training, authorized entrant training and authorized attendant training are all required. Documentation must be maintained on-site or readily available within 24- hours. Failure to provide requested information in the allotted time period will result in disciplinary action against the non-compliant contracted company.

Contractors must supply all of their own confined space entry / rescue equipment. You should never loan, lease or sell any confined space testing equipment, ventilation fans and retrieval / rescue equipment to contractors.
Prior to entering a confined space, all lockout, tag out and tryout procedures must be completed. Contractor must also notify confined space recuse team before entering and when exiting confined space. Initial atmospheric monitoring shall be performed prior to entering the space. The atmosphere shall be tested from the outside to determine if purging is necessary. All permit required confined spaces will require continuous monitoring. The results will be recorded on permit at a minimum of every 30 minutes or every time there is an atmospheric change.
While a confined space is occupied, an authorized attendant must be stationed outside the space at all times. Some form of communication must be

maintained between the entrants and attendants. This communication may be visual, radio, or other means of communication which will enable an immediate evacuation for any change in conditions which could reasonably impact the safety of the entrants.

Accidents in confined spaces present unique challenges and are often catastrophic. Injuries and fatalities involving confined spaces are frequent and often involve successive fatalities when would-be rescuers succumb to the same problem as the initial victim.

Crane Interferences Permit

The job of an overhead crane operator is very important. It requires strict adherence to rules and regulations that be must obeyed and responsibilities that must be accepted. Contractors need to be aware that all cranes are different and may have specific operating, safety, inspection and maintenance requirements. It is essential that they have the manufacturer's operating manuals and are familiar with your site-particular cranes.

Contractors should never operate any of your site overhead cranes except as approved by your Supervisor. Upon any approval, contractors are solely responsible for inspecting the equipment prior to operation. Contractors who operate cranes shall be trained and experienced. Proper documentation should be made available, indicating recent operating training or certification from a reputable institution. It is the responsibility of the operators to ensure the crane is in good operating order. Operators should never take loads over the heads of personnel or other persons. Likewise, personnel must never walk underneath loads.

A crane interference permit is to be completed whenever a job task has the potential to interfere with the operations of a crane. Contractors performing work near overhead crane (O.C.) operations (i.e., maintenance work being performed on a parked crane, personnel working from an elevated man-basket, the use of a mobile crane, floor repair activities, etc.) should comply with these procedures.

It is common practice to have a crane spotter positioned in a location that will provide them with good visibility and warning capability. Spotters shall have no other responsibilities, which may distract them from the spotter duties. An individual spotter can be used per individual crane to serve as a spotter for several contractors working in an area. The spotter is per crane for the activity that the crane may encounter. Crane signals should only be given by a trained and authorized contractor. They should use the standard signals when signaling the crane operator.

SCM, LLC

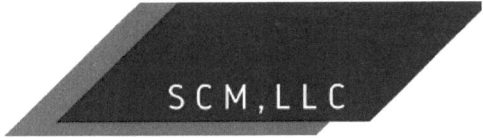

Sample Crane Interference Permit Application

DEPARTMENT: _____ DATE: _____

LOCATION OF WORK ACTIVITIES: _____

CRANE NUMBER: _____

EMPLOYEE/CONTRACTOR'S NAME: _____
EMPLOYEE/CONTRACTOR'S SUPERVISOR: _____

DESCRIPTION OF WORK TO BE DONE: _____
_____
_____

DATE AND TIME WORK TO BEGIN: _____
DATE AND TIME WORK TO END: _____

HAZARD CONTROLS: _____

NOTIFY SUPERVISOR AND CRANE OPERATOR
1. CAB, REMOTE, OR PENDANT CONTROLS TAGGED: _____ YES
2. CRANE OPERATORS NOTIFIED: _____ YES

CRANE OPERATORS NAME   CRANE OPERATORS SIGNATURE
_____   _____

*LOCKOUT OR RAIL STOPS ARE REQUIRED WHEN WORK IS TO BE
PERFORMED ABOVE THE ELEVATION OF THE CRANE RAILS

3. LOCKOUT/TAGOUT PROCEDURES IMPLEMENTED: _____YES _____ NO
4. TEMPORARY RAIL STOPS INSTALLED: _____ YES _____NO

5. CRANE SAFETY SPOTTER:
_____   _____
SPOTTERS NAME   SPOTTERS SIGNATURE

PERMIT ISSUED BY:_____

SAFETY SPOTTER RESPONSIBILITIES:

SAFETY SPOTTER CANNOT BE OPERATOR

MUST HAVE HORN AND RADIO: _____ COMPLETED

MUST REMAIN WITH OPERATOR/CONTROL BOX: _____ COMPLETED

EXPLAIN WORK TO BE PERFORMED: _____ COMPLETED

MAINTAIN SAFE DISTANCE FROM EQUIPMENT: _____ COMPLETED

IF SPOTTER LEAVES, JOB STOPS UNTIL RETURNS: _____ COMPLETED

CONTROLS TAGS REMOVED: _____ COMPLETED

LOCKOUT LOCKS OR RAIL STOPS REMOVED: _____ COMPLETED

JOB COMPLETED: _____ COMPLETED

Rail Safety Permit

Your contractors should possess a clear understanding of the hazards involved while working on or around train rails. Your company should refer to the Office of the National Rail Safety Regulator (ONRSR) guidelines when developing a procedure for all contractors working on your site. Your rail safety procedure should –

- Provide a description of your facility's rail infrastructure

- Identify the nature of the tasks and the location of work to be performed

- Identify credentials required to perform rail work

- Include a procedure for locking out rail sections to perform work, known as blue flag procedure.

Blue flag procedures are established to protect railroad workers, your teammates and contractors while working within six feet of the outside rail or less than twenty-two feet above the railway. In many cases, when working around an active railway system, you should lockout that section of railway and install a D-rail. Communication is critical when working around rails and a permit will be required before any work starts.

Always remember, only trained and authorized individuals can install D-rails, blue flags and operate rail switched. Even though there have been significant reductions in fatalities over the years, accidental deaths still occur.

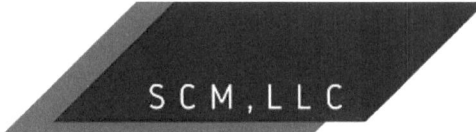

SCM, LLC

Sample Blue Flag Permit Application

DEPARTMENT: _____ DATE: _____

LOCATION OF WORK: _____

COMPANY: _____

JOB SUPERVISOR: _____

DESCRIPTION OF WORK: _____ _____
_____ _____

BEGINNING DATE & TIME: _____
ENDING DATE & TIME: _____

SAFETY CONTROLS:

1. IS WORK BEING PERFORMED ON OR LESS THAN 22 FEET ABOVE TRACK?
   _____ YES   _____ NO; IF YES, ARE D-RAILS AND
   PERSONAL LOCKS INSTALLED? _____ YES   _____ NO

2. IS WORK BEING PERFORMED LESS THAN 6 FEET FROM THE OUTSIDE RAIL?
   _____ YES   _____ NO

3. IS WORK BEING PERFORMED BETWEEN DUSK AND DAWN?
   _____ YES   _____ NO

4. ARE THE BLUE LIGHTS INSTALLED WHERE NECESSARY?
   _____ YES   _____ NO

5. HAVE THE RAIL YARD LEADMAN BEEN NOTIFIED?
   _____ YES   _____ NO

_____   _____
RAIL YARD LEADMAN NAME   RAIL YARD LEADMAN SIGNATURE

_____   _____
CONTRACTOR NAME   CONTRACTOR SIGNATURE

_____   _____
AREA SUPERVISOR NAME   AREA SUPERVISOR SIGNATURE

Walking and Working Surface

Ideally, workers should constantly be aware of their environment, particularly the walking and working surfaces. Unfortunately, other factors – personal matters, an upcoming meeting, and a conversation with a co-worker, can often affect one's concentration and attention to his or her surroundings.

Falls due to slips and trips are generally referred to as falls from the same level. Other types of falls often involve falling from or to a different level, either from an elevated area or through an opening in the walking surface.

Not all work places can be altered to prevent falls so there are several personal and environmental factors that contractors can keep in mind to remain safe at your facility. These factors include, but are not limited to:

- Using proper equipment, such as a ladder, if you must work or reach a higher level
- Walking at a safe pace, remaining alert to any obstacles that may be ahead; adjust pace and stride for the condition of the walking surface
- Wearing proper footwear for the job and work surface conditions
- Not carrying items that block the view ahead
- Not jumping from heights; instead, climb or ease down
- Promptly reporting any hazards
- Taking the time to pick up debris items or clean-up minor spills; never assume somebody else will take care of it
- Using personal protective equipment where necessary
- Closing all file and desk drawers immediately
- Using a flashlight to enter dark areas
- Storing heavy items down low; heavy items may be hard to handle on ladders or step stools

Floor System and Handrail Removal Permit

This procedure is designed to minimize slip, trip and fall (STF) hazards associated with the removal of floor gratings when performing maintenance or service tasks. Floor grating should not be removed without first contacting your facility's safety department and obtaining a completed Grating Removal Permit. Prior to grating removal and the start of work, the safety department will review the procedures and protective measures required by the permit.

This permit is also used when removing floor plates and handrails. Barricades isolating area must be installed prior to work beginning.

Preventing STF hazards can be a major challenge for any contractor or company. Using these measures will help reduce the probability of slips, trips, and falls. The ultimate responsibility, however, rests with the worker to remain alert, recognize hazards, and take the necessary steps to avoid being injured.

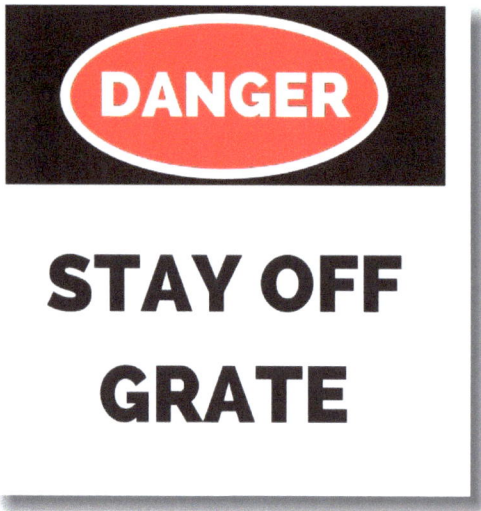

DANGER

STAY OFF GRATE

Excavation Permits

One of the most preventable hazards of construction work is the danger of trench cave-ins. Yet, every year in the United States, there are an estimated 75 to 200 deaths and more than 1,000 lost work days per year from trenching accidents. Other hazards associated with excavations and trenches include contact with numerous underground utilities, hazardous atmospheres, water accumulation, and collapse of adjacent structures. For these reasons, your company should have written procedures for all excavations performed at your site. Your policy should permit only trained and authorized personnel to create or work in excavations.

The purpose of this procedure is to protect personnel against cave-ins and other unexpected hazards associated with excavation and/or trenching operations. The individual authorizing such work will be responsible for forwarding all documents pertaining to the excavation work to the site's safety department upon completion of the work. Such documents may include, but are not limited, to confined space entry permits, opening dangerous systems permits, safety meeting minutes, and excavation audit reports.

Your facility's procedure for excavation work should apply to all contractors involved in performing work at your site. The individual authorizing such work is responsible for ensuring that the excavation/trench and related operations present no foreseen hazards and meet all requirements of this procedure as well as all applicable regulations. This individual also has the authority to promptly eliminate any or all foreseen hazards and keep, at a minimum, the following at the work site during excavation operations – the excavation permit, daily inspection checklist, and contractors work permit.

The purpose of these procedures is to ensure that excavations are conducted in a safe manner and in compliance with applicable regulations. The excavation procedure not only applies to the contractors at your site but also apply to company personnel such as area supervisors, utility locators, subcontractors, project and field construction managers, facilities engineer, and other competent persons and department heads. The procedure should clearly cover planning, performing, and closing out excavations in which contact with soil is expected – such as trenching, drilling, and removing soil – that meet any of these conditions at any time:

1. Depth is one foot or more

2. Power tools will be used

3. Utilities are identified

4. Any hazardous condition is likely to be encountered

In most facilities, the following operations are usually exempt from requiring an excavation permit: sampling soil, removing concrete/ asphalt and/or loading from stockpiles.

SCM, LLC

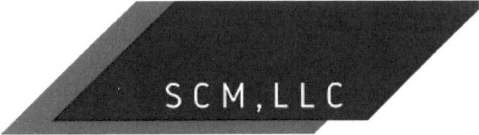

Sample Excavation Permit Application

Department of Requesting Permit: _____

Individual Requesting Permit: _____

Requested Date to begin excavation _____
Estimated Date of Completion _____

Purpose of excavation: New Construction   General Maintenance

Company Performing Excavation _____Individual Performing Excavation_____
Company Competent Person_____

Company Entering Excavation_____

Individuals Entering Excavation (List All)

_____   _____
Company Competent Person   Company Competent Person

_____   _____
Company Competent Person   Company Competent Person

Location of Excavation: _____

Identify Adjacent Public Utilities: (circle all applicable)
     Electrical   Telephone   Sewer   Water   Oxygen

     Natural Gas   Man Holes   Overhead Power lines

4

When Changes Are Needed

Be truly glad, there is wonderful joy ahead!

I Peter 1:6

Change Requests

Change requests have many different names, i.e. change order, scope change, project overrun, and cost variance. These different terms all essentially describe the same concept – a required change to the original plan. It is very important that this process is implemented in your facility/contractor procedures. Change requests can affect your facility safety, reliability, cost, project completion timeline, and effectiveness. For example, a blue print change request used to capture all new installations, upgrades, additions and modification made in or around your facility has to be converted to the associated engineered drawings to match change. This change can affect several of the factors previously mentioned.

Change requests typically originate from one of listed points:

- Newly installed equipment, structures, piping, valves, disconnects, etc.

- Problem reports that identify safety concerns

- System enhancement and upgrades

- Part number, pump size, cylinder bore, stroke, etc.

- Relocation of disconnects and emergency valves

- Removal of systems, machinery, piping, disconnect, etc.

Additionally, change requests may also originate from an operational or procedural stand point.

Upon completion, all change requests should be turned in to your engineering department. They should be submitted in the form of sketch or of an AutoCAD drawing. All associated information – such as location, material size, pipe size – should be included when turned in. Once change request and sketch are received, your facility drafter will modify the existing drawings to reflect the changes.

A change request form and procedure should also outline your policy with contractors. I would recommend not paying the contractor for his services until this information is received.

It is ideal to include your teammates and ensure that they follow this policy, as well all contractors.

SCM,LLC

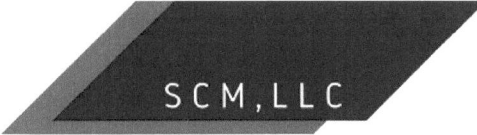

Sample Blueprint Change Request Form

Requester _____

# of Drawings Request _____   Date: _____

Engineering Dept Drawing Control Request #: _____
Job Order #: _____

Drawing Numbers: _____
Vendor: _____

TITLE BLOCK DESCRIPTION:

_____ UPDATE DRAWINGS   _____ CREATE NEW DRAWING
_____ VOID or COMPLETE REDRAW   _____ MARK-UP ATTACHED
_____ SKETCH ATTACHED   _____ EMAIL

DETAILS OF CHANGE REQUEST:
_____
_____

IF NEW DRAWINGS, LIST SOURCE OF DRAWINGS AND ATTACH VENDOR
TRANSMITTAL IF APPLICABLE.
_____
_____

EXISTING EQUIPMENT AFFECTED BY THIS CHANGE.
_____
_____

IF DELETING AN EXISTING DRAWING, STATE REASON:
_____
_____

_____ Check here for consultation with draftsman prior to drawing change

REQUESTED DATE OF COMPLETION: _____
PROJECT ENGINEER APPROVAL: _____ READY ISSUE _____
RETURN

FOR CORRCTIONS – SEE MARK-UP

REMARKS _____

_____

_____

(FOR ENGINEERING DEPARTMENTAL USE ONLY)

____ DRAFT PERSON   _____ DRAWING REPLACED IN HARD COPY FILES

_____ DRAWING REVISION COMPLETED   ____ 11 X 17 LIBRARY COPY

_____ CHANGES DOCUMENTED IN DATABASE

5

Getting The Job Done

*Few things fire up a person's commitment like dedication
to excellence.*

— John Maxwell

Important Factors for Working Efficiently

Set up your project for success, to accomplish this, you must ensure proper permits are in place and your contractors are crystal clear on work expectations. Additionally, it is important to address the following:

- **Communication** – regular meetings with your employees and contracted workers to share any safety concerns, project scope, and projected schedule

- **Support** – encouraging teamwork throughout the project. This is where people stop being individuals and become a part of an entity built by understanding the end result

- **Tools** – inspecting, testing, and checking all hand and power tools

- **Operator Training** – ensure all individual training is up to date

- **Rental Equipment** – schedule rental equipment, cranes and scaffolding as needed.

These important factors can help your project run smoothly if you thoughtfully execute plans with those concepts in mind.

Equipment Rental

When it comes to the type of projects that require contracted workers, many manufacturing industries have to obtain special equipment to have the work completed. Often times, the special equipment is rented. Many equipment rental companies are all about convenience and availability - making sure you have the right tool or equipment at the right time to get the job done. Those that regularly serve the manufacturing industry are familiar with the types of equipment normally used and usually have them readily available. Your company should have a policy that outlines rental equipment, safety during its use, the process for requesting equipment, evaluation of its effectiveness, and its cost.

The rental industry is all about convenience and availability - making sure you have the right tool or equipment at the right time to get the job done. Jobsite safety depends on understanding how to use your rented tool or machine before you put it to work.

Objective:

If your job requires the use of rented equipment, your company should have a policy that outlines rental equipment safety, request, evaluation, availability and cost.

Safety:

All rental equipment should only be operated by someone certified in using the equipment and should come equip with the following safety items:

- fully charged fire extinguisher (in the basket & body)

- operational safety manual

- flashlight

- back-up alarm

- legible control panel labeling

- legible safety decals

- toe plate completely around basket

- documented daily inspections

All operators of manlifts, scissor lifts, lull fork lifts, carry deck cranes, and other equipment must be certified by a third-party instructor before they are allowed to operate equipment.

Request & Evaluation:

All rental equipment requests must be made to your company's rental equipment coordinator by submitting a completed rental request form. Your request will automatically be forwarded to the rental Equipment coordinator for evaluation and processing. Once approved, your request will be scheduled for delivery of equipment.

When equipment arrives, you must check it out from the rental equipment coordinator to verify that all the required safety components are in place and working properly. The rental equipment coordinator will provide you with an equipment check out sheet.

Once the receiving inspection and check out sheet are completed, return the form to the rental equipment coordinator for filing and the equipment is now available for use. It is imperative that you are properly trained to operation the equipment you have rented. Remember, while using the equipment, you are 100% responsible for it as well as responsible for your safety and the safety of those around you. If your rental equipment needs any fuel or mechanic services, it is recommended that you contact your rental equipment coordinator for assistance. When the equipment is not being used, ensure that it is turned off, by using the E-stop button, to preserve the battery.

Daily or prior-to-use inspections are required on all rented equipment. These inspections must be documented and performed by the equipment operator. When you are done with your rented equipment, remove all trash, scrap metals, welding rods, etc. from the equipment. Return the equipment to the Rental Equipment Coordinator

Your original check out sheet will be used to check the equipment back in. Inspect the equipment and record any change in machine condition. Once completed, you are done with this equipment rental and no longer responsible for its use and condition. The Rental Equipment Coordinator will call the rental provider and schedule a pick up. Please keep your daily inspection sheets for your records. The check in and out sheets will be kept by the Rental Equipment Coordinator.

Cost:

Your rental equipment coordinator should have a list of preferred vendors from which to contact concerning equipment rentals. Usually this list is compiled through a bid process for cost effectiveness. In addition to cost, the rental equipment coordinator should also consider these items when renting –

- equipment reliability and equipment age

- mechanical support

- equipment safety items installed

- additional surcharge fees

- delivery and pick up costs

Material Removal Process

Many of your contractors are not based on your site. When projects require fabrication of steel or equipment, contractors may have to work from their

base shops. To complete projects requested by your company, materials must be delivered to the contractor's base shop.

The following procedures must be used when a contractor is removing materials or parts from your site to modify and return for installation.

Material Release Forms

Material release forms must be filled out and signed by an authorized contractor representative. These forms must also be signed by any site supervisor. The following requirements are industry-standard and can vary by facility –

- materials may be removed during daylight operation hours only;

- a completed material release form must be presented to the gate guard when departing the facility;

- the material release form is only good for the date specified;

- must include a complete inventory of all materials to be removed;

- the material release form must be completely signed by all parties;

- the name and clock/employee number of the individual(s) removing the material must be printed on the form;

- the facility's gate attendant should never let anyone leave your site with parts and/or materials without a completed material removal form.

Note: All leftover scrap metal from your materials should be properly placed in a container approved by the hosting company. This container should then be returned to the company once the job is complete.

Housekeeping

A worksite can be a place full of dangers. Practicing good housekeeping helps to reduce the risk of accidents in the workplace. Maintaining good housekeeping habits is one of the best ways to prevent certain accidents and illnesses from occurring while contractors are working on site. Keeping a list of best practices is a great way to keep track of all of the needed cleaning tasks to be accomplished every day.

Here are just a few industry-standard housekeeping do's and don'ts:

- **DO** gather up and remove debris to keep the work site orderly;

- **DON'T** permit rubbish to fall freely from any level of the project; use chutes or other approved devices to materials;

- **DO** plan for the adequate disposal of scrap, waste and surplus materials; designate areas for waste materials and provide containers;

- **DO** keep the work area and all equipment tidy and keep stairways, passageways, ladders, scaffold and gangways free of material, supplies and obstructions;

- **DO** secure loose or light material that is stored on roofs or on open floors;

- **DON'T** throw tools or other materials;

- **DO** keep materials at least 2m (5 ft.) from openings, roof edges, excavations or trenches;

- **DO** remove or bend over nails protruding from lumber;

- **DO** keep hoses, power cords, welding leads, etc. from laying in heavily travelled walkways or areas;

- **DON'T** raise or lower any tool or equipment by its own cable or supply hose;

- **DO** ensure structural openings are covered and protected adequately (e.g. sumps, shafts, floor openings, etc.)

6

More Important Factors

Make the most out of every opportunity.
Colossians 4:5

Observation Participation

A safety observation program can proactively prevent incidents and injuries through the monitoring, trending, and management of safe vs. unsafe behaviors. The effective communication of safe and unsafe behavior trends to the contractors and project management is critical to a successful program.
This program is intended to be used as a guide to all of your contractors. Each site may vary on their data collection system and how and what information is used and presented.

Your teammates will make observations of contractors at work out on the site. The observations will note the date and time, location, company, and number of employees observed. Each observation will be designated safe or unsafe, and if unsafe, further information and categorization of the unsafe act. Also, observers are trained to constructively correct unsafe behaviors and provide positive feedback on safe behaviors.

The observations should be input into a tracking database from which the behavioral trends of the workforce are determined. The results of the observations are communicated both to your teammates and on individual basis to the contractor companies. Focus areas from previous observations will be identified and emphasized to the workforce.

Safety Meetings

Conducting safety meetings is an effective way to instill a strong safety culture, avoid frivolous lawsuits and OSHA fines. It sends a clear message that safety is a priority. Developing a strong safety culture is important for maintaining an accident-free workplace.

OSHA requires contractors to conduct safety meetings on a regular basis and record the attendance. Regularly scheduled safety meetings allows the contractor to address important safety hazards and provide updates on recent incidents.

Research shows that safety meetings and safety programs reduce employee accidents. As a result, you can expect increased efficiency and reduced expenses with routine contractor's safety meetings. Ultimately, the contractors' safety meeting sends a clear message that safety is a priority.

Each meeting outline helps you comply with a specific OSHA safety training regulation, such as hazard communication, personal protective equipment (PPE), ladder & tool safety, lockout/tag out, material safety data sheets (MSDS), respiratory protection, machine guarding, etc. When planning your meeting, be sure to include topics that are relevant to the duties that are being performed in the workplace. It would be a waste of time to discuss hand tool safety if the use of such tools is unnecessary. Keeping the meetings on point will help you have more smooth and efficient safety meetings. Also, you should record the following information at each meeting:

- meeting date

- attendees' names

- safety and health issues discussed; include hazards involving tools, equipment, the work environment, and work practices

- recommendations for correcting hazards and reasonable deadlines for management to respond;

- name(s) of the individual(s) who will follow up on the recommendations;

- All other committee reports, evaluations, and recommendations

Your company should hold the following meetings and invite the appropriate level of contractors.

Daily tool box meetings:

Daily tool box or tail gate meetings should be held at the beginning of every shift change. The purpose of these meetings is to communicate expectations, inform and update everyone about the schedule of work to be completed that day. Any and all safety concerns should be brought up and discuss with a plan put into place to mitigate the safety concern. Most daily meetings will include operational employees and contractors working in a particular operational area. Daily tool box meetings normally last about 15 minutes each work day.

Weekly maintenance shutdown meetings:

Most Manufacturing Companies shut down there operation once a week to perform routine maintenance on equipment. Operational employees and contractors typically perform these routine maintenance task and checks. The purpose of meeting before this weekly shutdown is to plan the maintenance activities, coordinate personnel and define work expectations for everyone involved. Weekly shutdown meetings normally last a little longer than your daily meetings (thirty to forty-five minutes). Everyone performing or overseeing maintenance activities in that area should be in attendance.

Monthly contractors' safety meetings:

A regular schedule meeting with your core contractors is very important in contractor management. The purpose of this monthly meeting is to communicate a consistent message to all of your contractors at once. These meetings should address, at minimum, the following topics – safety concerns, environmental issues, policy updates, procedural implementations, training opportunities, and any current opportunities for improvement. All contractors working on your site or have regular schedule work at your facility should have a representative in attendance, such as the superintendent or site safety rep. The monthly contractors' safety meeting usually last between forty-five minutes and one hour.

Outage planning meetings:

Most manufacturing companies will schedule extended down-time to repair, upgrade or perform large scale maintenance on manufacturing equipment. Typically, these down time periods can last anywhere between five consecutive days to five consecutive months, depending on the amount of work that needs completed and the manpower available to perform the work. Before starting one of these extended down-time periods, you must meet internally and consider what projects will be schedule, the required manpower needed to perform them, and approximate amount of time needed to complete them. Upon determining this, you should schedule another meeting with all parties involved. The purpose of this outage planning meeting is to allow all operational employees, maintenance employees and contractors who are performing work during this extended shutdown period to completely understand and ask question about the job before it starts. This meeting will also give everyone in attendance the opportunity to hear what other jobs are going on in the area. With this information, they can proactively plan accordingly to complete assign projects with little or no interruptions in the time allotted. This type of meeting normally last about one to two hours. All personnel in charge of maintenance in that area should be in attendance along with safety representatives and contractor supervisors.

Annual contractor's owners meetings:

Many manufacturing companies will host annually a meeting with all contractors and vendors. These meeting are normally on the management and owner levels. The purpose of this meeting is to dial in on expectations from the top down on safety, cost, effectiveness, reliability and a general overview of current business conditions. Annual contractors owners meeting is normally scheduled months in advance to accommodate out of town owners who will be traveling to attend. A formal agenda should be in place to outline the meeting with at least one professional guest speaker relevant to your industry. You general manager and maintenance manager should also be on the agenda for opening remarks and closing remarks. These Annual Contractors Owners meeting normally last about three to four hours with lunch provided by the hosting company at the conclusion.

Emergency Drill Requirements

Your company should lead the charge in designing and conducting emergency evacuation and shelter-in-place drills at your facility. While contractor is performing work on your site, they should conduct routine drills and be aware of all emergency protocols.

Evacuation and shelter-in-place drills are scheduled throughout the year to ensure the readiness of the contractors in responding to any type of crisis that requires building occupants to evacuate a building or to seek protective shelter inside of a building. Evacuation and sheltering-in-place are the inverse of one another. You evacuate a building when the conditions inside the building present a hazard to human life, health or safety. If the conditions outside of a building presented a hazard to human life, health or safety, one would reverse the evacuation steps and shelter inside a building.

Evacuation Drills

Evacuation drills include alarm activation to ensure fire protection and reliability, along with an orderly, disciplined evacuation, followed by a thorough inspection of the building to immediately rectify any code related issues. Finally, an on-site discussion is held with contractors, staff and faculty to evaluate and improve, when necessary, the performance and efficacy of these drills.

Shelter-in-Place Drills

Shelter-in-place drills are a tactical response to a possible chemical, biological, radiological, nuclear or natural disaster. They are designed to provide a place of refuge for people and to give them a level of physical, emotional and mental comfort. During a shelter-in-place drill, building occupants are directed to pre-designated "shelter areas" within a building. This emergency drill is developed to assist facilities in planning, conducting and evaluating emergency drills at your facility. Emergency drills provide opportunities to practice emergency response and enhance the staff's ability to implement the facility emergency plan when it becomes necessary. The more familiar people are with something the better able they are to perform a task under pressure and in difficult circumstances. Emergency drills are important and beneficial in several ways including:

- Identifying weaknesses and deficits in emergency plan processes

- Identifying strengths in emergency plan processes

- Meeting conditions for coverage requirements

- Improving staff and patient readiness/preparedness levels

- Familiarizing staff and patients with the facility emergency plans.

Providing an opportunity for contractors and staff members to rehearse the actions they would take in a real emergency.

A Personal Experience in…

Safety Preparedness

One year, in preparation for one of our inspections with the Occupational Safety and Health Administration (OSHA), my company brought in a retired OSHA inspector to perform pre-audits with our base contractors. This guy was like a walking OSHA manual! He knew every regulation to a T. I accompanied him in inspecting our contractors and was absolutely amazed at his knowledge. We inspected safety documents, workshops, office trailers, and every tool belonging to our contractors. We also performed one on one interviews with contractors on all levels of experience and responsibility. After inspections were complete, he provided us with a detailed closing meeting with all nonconformance issues outlined. This pre-audit inspection was a beneficial tool in showing our contractor what to expect with the actual inspection. It also gave us an opportunity to correct issues found prior to that inspection with OSHA taking place.

Other than his keen knowledge and commitment to safety in the workplace, there was a simply phrase he shared that I will never forget – you can't see the tree because of the forest. *It took me a moment before I finally understood what he was saying. In essence, he was saying don't focus on the big picture; instead focus on the smaller, more manageable issues to achieve the overall goal.*

7

When the Job is Done

All hard work brings profit, but mere talk leads only to poverty.

Proverbs 14:23

Reporting Contractor Incidents

Managing incident reporting and investigation in your organization is critical and very important. Conducting thorough incident investigations and root cause analyses to reduce risk, avoid future incidents and drive continuous improvement will benefit your contractors, your team, and your company. The basic incident reporting process has become significantly easier. Companies are able to actually monitor whether people are following through on critical steps in the reporting process.

Incidents are unplanned events or occurrence that may cause damage to equipment, harm or be fatal to the human body and have a direct impact on the environment. There are three primary reasons an incident occurs – (1) personal responsibility; (2) defective equipment; and (3) inadequate procedures.

Incident investigation is the official examination of an unplanned event that caused damage to something, injury to the human body or directly negatively impacted the environment. Incident investigations are normally performed by area supervisors and your safety department designated individual. Incident reporting is critical in the prevention of future incidents. No matter how small the incident, it must always be reported, investigated and communicated to everyone.

Root cause analysis is a useful process for understanding and solving incidents. As an analytical tool, root cause analysis is designed to help identify not only what and how an incident occurred, but also why it happened. With the findings of your Root Cause Analysis, you should be able to implement preventions to insure the incident never occurs again.

Incident reporting, investigation and root cause analysis programs will benefit your company in many ways:

- **Prevention:**
  Reporting incident even when nothing happen, no damages occurred and no one was injured is key when it comes to preventing future incidents

- **Investigation:**
  Proper incident investigation will help you determine the proper root cause which normally is, Personal responsibility, defective equipment or inadequate procedures

- **Communication:**
  Incident communication within your site, company and industry is very important. Sharing and receiving incidents from other facilities can proactively prevent reoccurring incidents

- **Risk Reduction:**
  Based on your past incidents your company can trend the data to analyze possibilities of future incidents and implement preventative measure to reduce risk

- **Elimination:**

Identify the root cause and eliminate it from your work area or develop a procedure to safely work around it.

Vehicle Incidents

All contractors' vehicle incidents at work, while traveling to work or home from work must be reported, investigated, recorded and the findings communicated. These incidents should also be trended to determine commonalty. Once results are made, a programs should be implemented to support safe vehicle driving practices at all times.

Environmental Incidents

If incidents result from spills of oil, gas, chemicals, or any foreign product occurs, you should immediately contact the proper safety and environmental individuals at your site. An incident investigation report must be completed and the findings communicated. In some cases, your report may be viewed by the EPA or other outside agencies.

Conducting Contractor Training

It is important for your employees to understand the nature of the relationship between your company and the contracted employees. The desire to treat the contractor as part of the team is understandable, but your employees must realize the potential risks they impose on your company and the contractor if they allow special treatment in any number of situations. Just as different rules apply to the visitors at your site, there are, too, rules for contractors. Your employees must be aware of these rules, regulations and policies during all ongoing contractor work in his or her area. It is also equally important that your company assigns someone as a key contact for contractors. This person should be knowledgeable in all related contractor programs, policies and procedures. This individual will work closely with many different departments to ensure all requirements are met before, during, and after contractors work on your site. It is very important that this individual provide contractor training for your employees as well. This can be done once a month with a different contractor related topic during your supervisors meeting.

When dealing with contractors, there are some very important points to remember:

- Contractor personnel are not your employees. However, contracted personnel are no different than you and me. They should be treated with the same respect as you do employees of the company. as if your weekly pay is depending on their performance;

- Identify contractor personnel as such with distinctive badges and proper PPE (such as hard hats)

- Clearly identify the contractor's work area. This will help preclude any appearance of a personal service relationship between your employees and contractor personnel

- Contractor identification should also extend to email accounts. Email addresses and signature blocks should clearly identify contractor personnel

- Respect the employer-employee relationship between contractor and their employees

- Be aware of intellectual property rights in the federal workplace. The terms of the specific contract will determine the contractor's rights

- Report possible conflicts by contractor personnel to include violations of the law (including but not limited to Procurement Integrity statutes and regulations). Be sensitive to appearances created by close relationships between your company and contractor personnel

- Safeguard proprietary, Privacy Act, and other sensitive and nonpublic information. Release of certain types of information to unauthorized contractor personnel could violate the Procurement Integrity Act, the Trade Secrets Act, the Privacy Act, the Joint Ethics Regulation and/or other laws that could subject the releaser to civil and/or criminal penalties

- Clearly describe all contract tasks and utilize the contractors work permit to document

- Ensure only the contractor's task leader assigns jobs to contractor personnel however, provide instant feedback to all contractors when it comes to safety concerns, good or bad

- Lead by Example, Remember, Orders and commands don't plant the seed of commitment, you leadership does

- Contractors are employed for the same reason we are – to support their families. They truly have no intent of having an unsuccessful day. It is our responsibility to insure they have all the information needed to complete every job safely and on schedule

There are also a few things you want to steer clear from, as well, while working with contractors. Below is just a short list of contractor don'ts your organization and its employees should always consider regarding your relationship with your contracted workers:

- **DON'T** become involved with the contractor such that your judgment alone forms the basis for contractor actions such as – selecting contractor personnel, directing and scheduling individual contractor tasks on a continuous basis, supervising contractor personnel, rating an individual contractor's personnel performance, hiring or firing individual contractors, determining who should perform contract tasks, pressuring the contractor to use "favorite" personnel, or insisting on particular personnel actions

- **DON'T** use your company and contractor personnel interchangeably

- **DON'T** intervene in the contractor's chain of command

- **DON'T** require "out of scope" work, personal services, or favors during contracted work

- **DON'T** give any contractors a competitive edge outside of the contract or bid package

- **DON'T** solicit or accept gifts from contractor personnel. The rules for giving and getting gifts are outlined in the contract between your company and the contractor

- **DON'T** pay for contracted services until you 100% satisfied (expectation complete, safety items closed, Housekeeping brought back to normal, all permits closed out, etc.).

Evaluation and Feedback

All of your contractors should consistently receive feedback on their performance by your employees. It could be as simple as having a conversation with them about the work being or by thanking them for doing a good job.

Evaluations shall be performed in an objective, consistent, and well-documented manner. Your company should have a standard form for contractor evaluations. Many companies now have web-based evaluation tools in place for contractors. However, a simple word document with a rating scale and the evaluation subject areas should suffice. (See sample form).

It is important to provide an overall performance evaluation on all of your contractors annually. Doing so helps to promotes continuous improvements and reinforces best practices with your contractors. This evaluation should be completed by operational and maintenance employees that works with contracting companies at your facility. Normally, you should evaluate each company in five main performance areas – safety, cost, reliability, quality and effectiveness. These areas should be rate on a scale from 1 to 5, with 1 being poor performance and 5 being outstanding performance. Your employees should rate each company performance based on their work history over the past year. I would also add a comment section to capture any additional comments the evaluator may have. Once you have received all evaluation forms back, you should collectively average each contractor companies overall score in each performance area and summarize the comments per company.

This information is vital to your contractors and can determine future direction and relationship with your company so, formally set up a meeting with each contracting company supervisor. A recommend thirty minutes per company to cover the entire evaluation process, results and additional comments is ideal.

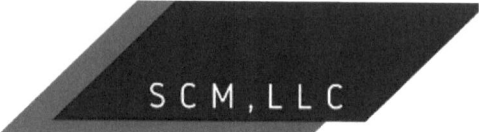

Sample Contractor Evaluation Form

Contractor Company: _____

All Contractors were evaluated on their work performance. The rating scale is as follows:

5 - Outstanding
4 - Above Average
3 - Average
2 - Below Average
1 - Poor

The chart below is an average of these results from supervisors and lead persons of various areas. Please see the chart below for your performance.

Areas of Evaluation	Performance Rating
Safety	
Cost	
Efficiency	
Reliability	
Quality	

Comments:

_____
_____
_____
_____
_____
_____
_____

8

Going Above and Beyond

Tomorrow…your reward for working safely today.
— Robert Pelton

Federal and State Recognition Programs

Many government agency offers voluntary and cooperative programs focused on reducing injuries, illnesses, and fatalities. These programs offers on-site consultation services that help struggling contractors improve with their safety performance. Contracting companies that work above and beyond the state and federal requirements to ensure that safety is a top priority are often recognized for their achievements. There are several different programs that promotes a philosophy that all associates of an organization strive for, on a voluntary basis, to ensure a safe workplace. Programs like these have many benefits but, ultimately the workers who, at the end of day, arrive home safely and healthy to their families who need not fear the tragedy of worker death, injury, and illness receive the greatest benefit.

Please encourage your organization and your contractors to participate in programs like these. A couple of such programs are listed below:

Safety & Health Achievement Recognition Program (SHARP) – This program recognizes small companies who used OSHA on-site consultation program services and operate an exemplary injury and illness prevention program

Voluntary Protection Plan (VPP) – This program recognizes companies who have implemented effective safety and health management system and maintained injury and illness rate below national averages for their respective industries.

Personal Responsibility & Incident Prevention

Personal responsibility is defined as a person's "response-ability," that is, the ability of a person to maturely respond to the various challenges and circumstances of life.

Taking personal responsibility also means that when individuals fail to meet expected standards, they do not look around for some factor outside themselves to blame. Personal responsibility in your company must start with you. Like any supervisor or leader, you must lead by example and practice what you preach.

Accident prevention is a major component of in the operation of an effective organization. Preventing accidents in the workplace should be always a core focus area. You should invest a significant amount of resources on this area (1) to train personnel, (2) to ensure the selection of proper tools, and (3) to execute safety perfection.

Identifying risks and seeking prevention measures is everyone's responsibility. This type of commitment is best illustrated by empowering your employees. Being part of the solution will guarantee buy in and keep your employees involved.

Contractor Community Involvement

The workplace is the ideal environment to promote volunteerism, whether activities involve the entire team or simply encouragement for individuals to engage in the community.

Community involvement is an integral part of creating a positive workplace. Companies and contractors committed to the area where their employees live and work emphasize giving back. Contractors that encourage community involvement distinguish themselves from their competitors, and see many benefits, including loyal customers and happier employees. For example, many companies provide employees with time to participate in programs to help area nonprofits, particularly during the holiday season. It would be beneficial to implement some sort of community involvement imitative within your organization. Some benefits of your company's participation in community service include relationship building within your working community, increasing employees' opportunity to participate in individual giving, creating a custom volunteer plan where employees/contractors can volunteer whenever needed, and providing evidence to your customers regarding your organization's commitment to community. Often times, owner companies will acknowledge contractors that understand the importance of this connection between good community stewardship and good business.

Whatever measure you use in giving - large or small - it will be used to measure what is given back to you.

LUKE 6:38 (NLT)

In Conclusion

Successful Contractor Management, LLC (SCM) is a full service agency with a strong focus on contract individuals and companies. With more than twenty years in this industry, we have realized the need for the successful management of contracted services. It is imperative that, in order to maintain successful contracted services, there are several components that should be considered – safety, procurement, work procedures, training records, self-inspection and communication. Onsite or offsite, locally or internationally, contractors are dedicated to providing service to help your company succeed. Contractors are all sorts of people, some you may know as family, friends, neighbors, church members, golf buddies, etc. They work day in and day out with one ultimate goal in mind – to take care of their families. This is the single, most sought after goal by everyone in the workforce today. Contractors are no different than you and me. While working on any job site, they should be given the highest level of treatment, attention, care and guidance. In fact, they should be held to higher standards because they may not be as familiar with your area, equipment or process. Often times, when a contractor has an incident, one of the primary reasons is lack of knowledge. When asked, the answers are usually "I didn't know …". This lack of knowledge comes from lack of communication prior to contractors working on a new job site or in a new area.

When coordinating contracted work, you must plan according, meet consistently, over communicate and execute carefully. Every contracted job, no matter how big or small, should receive your full and undivided attention from start to end. I encourage you and your company to develop and implement a robust contractor management plan as soon as possible, if you do not have one. This plan should include all of your contractor policies, procedures, permits, and forms. These permits and forms are very important tools. They will be meaningless, however, without first focusing on personal responsibility and prevention. As we all know, safety is not a policy, procedure or some gadget. It is a state of mind.

As a leader, you must motivate and encourage your contractors to focus on personal basic safety on and off the job. Negative behaviors, short tempers, and bad attitudes away from work will show up in due time and could result in a negative outcome. These personal issues must be identified and rectified before starting work.

Keep in mind that your contractor success is your success. Always set them up to succeed. Failure is not an option. When conveying this message to your contractors, please remember this – orders and commands do not plant the seed of commitment; leadership dose.

This book, this course and a comprehensive consultation is designed to help you, your company and all contractors work more efficient, effective and safe. These lessons will also result in quicker contractor approvals, lower cost and on time completion.

Thank you.

Glossary of Terms Used

area maintenance supervisor – an individual responsible for all maintenance requests and for the preventative maintenance in one particular area of a facility/plant

area operation supervisor – an individual responsible for the operation and production in one particular area of a facility/ plant

blood borne pathogens – infectious microorganisms in human blood that can cause disease in humans

blue flag permit – blue signals which prohibit movement of a train through the work area used when work is being performed directly on, over, or under railroad tracks

blueprint – a design plan or other technical drawing

blueprint change request – When a drawing revision or **change** notice has been prepared, the drawing usually consists of red line mark-ups, new drawings or a sketch attached.

Certified Professional Engineer – an engineer licensed and certified by a state board of registration to practice engineering; only a certified professional engineer may prepare, sign, seal and submit engineering plans and drawings to a public authority for approval.

competent person – an employee who is able to recognize hazards associated with a particular task, and has the ability to mitigate those hazards. Many OSHA construction standards require someone onsite – such as a foreman, supervisor or other employee – to be designated as a **competent person (Training and authority).**

confined space – a space that has limited or restricted means for entry or exit and is not designed for continuous occupancy

contractor – a person or company that undertakes a contract to provide materials or labor to perform a service or do a job for your company.

contractor profile sheet – a document designed to capture contractor-specific company information, such as address, work classification, phone number, etc.

contractor evaluation – an activity used to assess the contractor's safety, technical and programmatic progress, approaches, and deliverables

contractor incident – any work-related incident resulting in a contractor employee injury or near miss

Contractor Safety Summit – An annual meeting with contracting company owners and corporate safety representative

crane interference permit – a permit used to prevent overhead cranes from contacting mobile equipment while working in a building with an active crane runway

DCR – drawing change request

daily contact – an individual assigned to oversee contracted work in that area or person who requested contracted work to be done

down day meeting – a meeting scheduled to plan maintenance work during a 12 or 24 hour planned operating unit shutdown period.

electronic maintenance software – a web-based system for inventory management, work order tracking, and maintenance management system.

emergency plan – a course of action developed to mitigate the damage of potential events that could endanger an organization's ability to function

emergency evacuation – the immediate and urgent movement of people away from the threat or actual occurrence of a hazard.

excavation – the process of moving earth, rock or other materials with tools, equipment or explosives, usually to allow the construction of a foundation.

fall protection – any means used to protect workers from falls from heights

fire watch – a person assigned to observe ongoing hot work to identify and react to hazards

fire prevention plan – a plan put in place to prevent a fire from occurring in a workplace

floor grating – industrial flooring assembled in a grid, used to cover or floor any various openings; it provides a strong and durable surface but allows air, light, heat, sound and water to pass through. It is strong and durable, and virtually maintenance free

hazard communication – recognition of dangerous materials in their work environment and the hazards these materials present.

hot work – any process that can be a source of ignition when flammable material is present or can be a fire hazard regardless of the presence of flammable material in the workplace. Common hot work processes are welding, soldering, cutting and brazing

housekeeping inspection – the contractor's responsibility that job site must be as clean and orderly as possible while work is being performed. This inspection should be completed by the contractor constantly and periodically by your company.

IBC – International Building Code; a comprehensive system of codes that address the design and installation of building systems through requirements that emphasize performance

Independent Contractors Agreement – a legal document signed by your company and the contractor that governs the relationship between a company and contractor.

incident investigation – the account and analysis of an incident based on information gathered by a thorough examination of all contributing factors

in-house project – a project that is executed within your company

in-place shelter- is the use of a structure and its indoor atmosphere to temporarily separate individuals from a hazardous outdoor atmosphere(Tornados, Hurricanes, heavy storms, etc…)

invoice – a list of goods sent or services provided with a statement of the sum due for these goods and/or services; a bill

material removal form – a document used to approve removal of material, supply, parts or any good belonging to your company form your site

MEWP – Mobile Elevating Work Platform

OSHA – Occupational Safety and Health Administration

OSHA 300 log – a listing of all injuries and illnesses at your jobsite required by OSHA, to be maintained by a facility/plant

observation program – a way of gathering data by watching behavior, events, or noting physical characteristics in their natural

occupational noise exposure – industrial noise, or **occupational noise**, is often a term used in relation to environmental health and safety, rather than nuisance, as sustained exposure can

outage planning meeting – a meeting to plan upcoming schedule down time and maintenance/ new installation with contractors

personal protective equipment – protective clothing and/or equipment – including, but not limited to, helmets, goggles – designed to protect the wearer's body from injury or infection.

permitted projects – all projects requiring county or state permits before starting.

project – a temporary endeavor undertaken to create a unique product, service or result.

purchase order – a commercial document and first official offer issued by a buyer to a seller, indicating types, quantities, and agreed prices for products or services

quote – the estimated price of a job or service

rail stops – mechanical devices design to stop a crane traveling on a rail system

risk assessment – a systematic process of evaluating the potential risks that may be involved in a projected activity or undertaking

roof access permit – a form that must be fill out and approved before anyone can access any roof areas at your facility

root cause – an initiating cause of either a condition or a causal chain that leads to an outcome or effect of interest.

SCM – Successful Contractor Management

STF - stains, trips and falls

safety meeting – a formal or informal group meeting designed to get employees actively involved, encourage safety awareness and motivate contractors.

safety spotter – an individual who acts as an additional set of eyes watching out for you while you perform a specific task

safety orientation – the process of introducing new, inexperienced, and transferred contractors to your site. This training is prepared to make them aware of health, environmental and safety expectations

shelter-in-place drill – a practice run for a natural disaster, such as a tornado or earthquake

subcontractor – an individual person who is hired by a general contractor (or prime contractor, or main contractor) to perform a specific task as part of the overall project and is normally paid for services provided to the project by the originating general contractor

tool box meeting – an informal group discussion that focuses on a safety, progress and any topic of concern often held daily to promote your contractor's safety culture

VPP – Voluntary Protection Plan

walking working surface – every floor, working place and passageway

web-based data management system – a system that houses contractor-related information – including, but not limited to, procedures, employees training, certifications – for public viewing and access

work order – an order received by the contractor from the customer that specifically outlines project details, cost, and management approval

work permit – an in-house document created to ensure contracted work is performed in accordance with safety procedure and communicated to everyone involved

Successful Contractor Management, LLC (SCM, LLC) is designed to change your contractor relationship and reinforce your overall safety culture. This book was written to rehabilitate the lives of everyone working with contractors in our industry. Having personally worked as a contractor and then in a role in which I managed contractors in the manufacturing industry for over twenty years, I have develop a very deep passion for successful and safe contracted work. After reading this book, I hope you have learned a variety of personal skills, procedures, policies and processes to use in managing contractors. Within the book, I have also share many best practices used in large manufacturing industries. Many of these best practices have been viewed by federal and state agencies as best in class experience with contractors. It is my belief that everyone has the right to go home after work the same way they came – uninjured and in good health. After reading this book, we are confident that you will develop a safer and more efficient relationship with your contractors. You and your company will benefit from the experience and services of SCM, LLC and raise to new heights in contractor excellence and extending the right to go home the same way they came to your contractors.

Bernard is currently the Contractor Coordinator for major steel manufacturer in rural South Carolina. He has established over nineteen years' experience specializing in contractor management. His most noteworthy accomplishments include the development of hundreds of contractor programs, policies and procedures, overseeing over thirty-eight OSHA Voluntary Protection Plan (VPP) inspections with industry contractors. Bernard also provides contractor management training and mentorship to his organization's subsidiaries, major contracting companies, and local manufacturers. He is also a proud veteran of the United States Navy and holds professional industry certifications from North Carolina State University.

Therefore everyone who hears these words of mine and put them into practice is like a wise man who built his house on the rock.

MATTHEW 7:24-25 (NIV)